I0467926

BE THE CHANGE: SAVING THE WORLD WITH CITIZEN SCIENCE

2ND EDITION

CHANDRA CLARKE

Copyright © 2013-2017 Chandra Clarke All rights reserved.

ISBN: 1500595500
ISBN-13: 978-1500595500

Without limiting the rights under the copyright reserved above, no part of this publication may be reproduced, stored in, or introduced into a retrieval system, or transmitted in any form or by any means (electronic, mechanical, photocopying, recording, or otherwise) without prior written permission.

Paperback Edition, License Notes
While every effort has been made to ensure the accuracy and legitimacy of the references, referrals, and links (collectively "Links") presented in this book, the author is not responsible or liable for broken Links or missing or fallacious information at the Links. Any Links in this book to a specific product, process, web site, or service do not constitute or imply an endorsement by the author of same, or its producer or provider. The views and opinions contained at any Links do not necessarily express or reflect those of the author.

CONTENTS

HOW TO USE THIS BOOK

This book will introduce you to the wonderful, amazing, and exciting world of citizen science.

This is a world in which it is possible to go on a wildlife survey in a national park, install software programs on your computer to find a cure for cancer, have your smartphone log the sound pollution in your city, transcribe ancient Greek scrolls, or sift through the dirt from a site where a mastodon died 11,000 years ago—even if you never finished high school.

This book assumes that you are new to the movement and might not know too much about it yet. It is divided into two parts.

In the first part, I talk about citizen science as a concept: I explain what it is, how important it is, and why I believe we need much, much more of it. I then go on to talk about how it can personally benefit you, how you can get involved, and what it might mean to you if you did.

I encourage you to read Part I in full, so you have some background and context that will make your involvement more meaningful. However, if you're keen to jump right in, then head to Part II, where I have assembled a big list of projects that you can start right now. I have organized these by the level of involvement required, starting from the very basic "donate level" to the more complex (and thus much more interesting!) hands-on projects.

One note of caution: because the listings here describe projects, you may run into one or two that have been wrapped up by the time you read this. If you find one like that, please do let me know so I can update the book. (http://www.citizensciencecenter.com/get-in-touch/)

Meanwhile, you can also head over to my site, www.citizensciencecenter.com, and sign up for the newsletter. It alerts you to new projects as they come up, and it's completely free.

PART I — ALL ABOUT CITIZEN SCIENCE

So just what is citizen science, anyway?

Roughly speaking, citizen science is research that is conducted by non-professionals, or "amateurs"—ordinary people like you and me. It is known by a few different names; you might hear it referred to as participatory science, networked science, or crowdsourced science.

Citizen science has a couple of defining characteristics. The first is that if you decide to get involved in a project, most of the time you will be acting as a *researcher*, and not a *research subject*. You won't be asked to take a drug to see how it works, go through behavioral tests, or sit in a simulation. (That sort of thing is usually known as volunteering for a study, or participatory research.)

The second characteristic is that citizen science projects are usually organized by professional researchers and involve groups of participants, large or small. In that sense, citizen science is different from what has been called *amateur science*, which typically involves unpaid individuals working away on something in their spare time.

Citizen science has actually been around a long time. In the United States, the Audubon Society has been running annual bird counts at Christmastime since 1900.[1] Various astronomical societies have been making and pooling their starry observations ever since telescopes became relatively cheap and easy to mass produce.[2]

Even before these group efforts, there was a time when all science was done by "amateurs." Consider Charles Darwin[3]

7

or Percival Lowell,[4] both of whom used their own resources to pursue their passions (zoology and astronomy, respectively), but who didn't initially work for universities or research institutions. They were not what we think of today as professional scientists. Then there's Gregor Mendel,[5] the monk widely considered the father of genetics; Mary Elizabeth Barber,[6] who made important contributions to South African botany and ornithology; and Eva Ekeblad,[7] who discovered how to make flour and alcohol out of potatoes. In many ways, citizen science is a return to the way science used to be done: by enthusiastic, committed people just like you.

At this point, you might be wondering: if it has been around so long . . .

Why am I just hearing about it now?

There are several reasons why the citizen science movement has really taken off in recent years. Let's look at a few of them.

We have the technology

For starters, you can thank the web for increased access to a wealth of exciting new opportunities that benefit us all. The sudden explosion of citizen science projects stems directly from the Internet.

The Internet has dropped the cost of communication to almost zero. Previously, if researchers wanted to recruit people for projects, they would have had to place an ad in a publication and hope that enough people would see it to make the cost worthwhile. Communicating with participants would have required making hard copies of memos or newsletters, stuffing them in envelopes, and paying yet more money to send them via the postal service. Now, of course, you can put up a basic website using free software in less than a day, with website hosting costs as low as $10 per month,[8] and the cost of a basic Internet connection (for sending emails and updating the website) as low as $40[9] per month.

The cost of developing the "back end" of websites to track, process, and store large amounts of data has dropped, too. Hardware such as computer processing power and memory has become incredibly cheap. According to a chart compiled by John C. McCallum,[10] the price of a single megabyte of memory would have been more than US$400 million in 1957. Today, the same megabyte costs less than half a cent. Much of the software used to power websites is free and open source[11] (for example, Apache, WordPress, MySQL).

Indeed, even the cost associated with the most expensive aspect of setting up the more technologically complicated citizen science projects—the required human resources— is dropping, as more and more people around the world are gaining and offering the IT development skills needed. This is true of both web development and the more specialized "app" development. At oDesk, a source for hiring freelance programmers, more than 1,700 web programmers are available at rates of less than $10/hour.[12]

We have the data

This same technological price drop means that researchers are able to collect vast amounts of data. Collections that have lingered unexamined for decades in the dusty archives of museums are now being digitized at a

9

staggering rate.[13] Sensors that measure a wide variety of variables, such as temperature, wind speed, and the light spectrum, are getting cheaper and easier to acquire,[14] and they can be deployed on any number of stationary, mobile, or robotic platforms. Apps allow citizen science project designers to access a huge network of mobile sensors on phones and tablets. In the United States alone, by 2013, 34% of adults had a tablet,[15] and 56% of adults had smartphones.[16]

This means that researchers can now gather more data on any subject than any one person could hope to review in a lifetime—in some cases, there is more data than could be reviewed over several lifetimes. The famous Hubble telescope has produced more than a million observations[17] so far.[18] In the Mapper Project,[19] a citizen science initiative that set out to analyze images of Pavilion Lake in Canada, participants analyzed a whopping 1,113,706 images and submitted 1,481,526 classifications. In all, project members made 11,561 comments about what they saw underwater, and contributed more than 2,227 person-hours of work. To achieve the same results, a single researcher attempting the project alone would have had to work flat out on *nothing* but image classification for more than a year.[20]

You might be wondering why these same wonderful technologies can't be used to process collected data; in other words, why is human intervention required at all? Well, in many cases, these technologies *are* used to process that data. Cloud computing, for instance, is being used to model protein-to-protein interactions and to manage large watershed systems.[21] However, humans can still do many things—such as image recognition and classification—far better than computers—for now, anyway.[22]

We have the brain space

As a society, at least as far as the North American middle class is concerned, we like to claim that we are constantly busy: we're overworked, overstimulated, and overscheduled. That may be true for some people, but the numbers suggest that we actually have far more free time than we like to admit.

Clay Shirky talks about a concept called "cognitive surplus." In his book of the same title,[23] he talks about how modern life has provided us with astonishing amounts of free time. For example, he suggests that Americans (just Americans!) watch more than 200 *billion* hours of television every year.[24] If that statistic doesn't boggle your mind, consider the online game *World of Warcraft*: one source[25] suggests its 11 million players have spent as much as 5.9 million years on it . . . or as much time as humanity has spent evolving as a species.

It's not all just fun and games, either. The Corporation for National and Community Service suggests Americans formally volunteered more than 8.1 billion hours of their time in 2010. Shirky believes more than 100 million hours[26] have been donated to producing and refining *Wikipedia*, perhaps the most famous example of crowdsourced and collaborative work. Meanwhile, more than 100 hours of video are uploaded to YouTube every minute of every day.[27]

The numbers also suggest people may be looking for ways to do something meaningful with all this free time. Looking at web search data,[28] phrases like "what is the meaning of life?" and "finding your purpose" consistently trend at the high end of the scale—at least for those people who use Google as an all-knowing oracle. The question "What is my purpose in life?" gets at least 1,000 searches a month on Google, worldwide.[29]

We have the budget cuts

Although research and development funding as a percentage of the GDP remained relatively stable up until 2009,[30] the global financial crisis has had a major impact in many countries, including Greece,[31] Spain,[32] and, to a lesser extent, the UK.[33]

Even without financial crises, science funding is a contentious issue at budget time, and research and development funding seems to get a disproportionate[34] amount of scrutiny. Perhaps this is because certain aspects of science spending (for example, NASA in the United States) are highly visible, or perhaps because the benefits of research spending are so difficult to quantify directly.

Either way, individual studies can gain or lose funding depending on the grant-writing skills of the researcher, the popularity of the topic being studied, local and national politics, and a variety of other factors. As a result, most research projects have to operate within very tight budgets.

Because most citizen science projects rely on volunteers, citizen science represents a way for researchers to get a lot of work done in a very low-cost way. This, in turn, speeds up the pace of the research, discoveries, and progress.

Annnnd . . . we also have big problems

The daily headlines can make for some depressing reading.

Humanity has made great leaps forward on some issues. The rates for polio, for example, have plummeted from an estimated 350,000 cases in 1988 to just 223 reported cases in 2012.[35] On other fronts, we're not doing so hot. Diarrheal diseases, which are completely preventable, remain one of the top 10 causes of death every year.[36]

Other diseases seem even more intractable. Cancer remains a frightening and complex beast, with some types (like stomach cancer) declining, and others (like trachea, lung, and bronchus cancers) increasing sharply.[37] And the treatment options remain pretty much the same as they have for decades: radical surgery, radiation, and chemotherapy. We have better targeting with these approaches, but they all still leave the survivors with harsh side effects and major quality-of-life issues.

Meanwhile, climate change has become one of the defining problems of our era. The science involved in figuring out climate change is difficult, and the detractors are loud and, in some cases, well-funded. To make matters worse, major geopolitical and economic barriers stand in the way of solutions.

All of the above can make an individual feel rather helpless. What can you, just one person, do? Sure, you can donate to an environmental organization, but you may wonder how much of your money gets spent on solutions versus administration or other expenses. You can drive a more efficient car and recycle, but you probably wonder what difference it makes when you see trees being bulldozed for a development, or read about the record number of car owners in China.

At the risk of sounding like a motivational poster, this is where citizen science—and snowflakes—come in. As Vesta Kelly is credited with saying, "Snowflakes are one of nature's most fragile things, but just look what they can do when they stick together."

Why do I need citizen science?

Finding the time—even for a really good cause—can be difficult. After all, you've got bills to pay, perhaps some kids to raise, and probably a lot of other things that demand your time and attention.

It turns out, however, that doing citizen science can be good for you personally. Here's how:

You can relax

They're called "guilty pleasures" for a reason: those things we do that we know aren't particularly enriching or good for us. Perhaps they're even "bad" for us! A guilty pleasure might be something like reading a gossipy celebrity magazine, going for that extra beer, or getting hooked on a terrible TV series. (For me, it's ridiculously unbelievable action movies and red wine . . . often at the same time.)

But the neat thing about citizen science projects is that many are designed to yield the same types of fun and rewards as those generated by some of our pastimes— without the guilt. Several projects are designed with the same attributes as popular games. If you like *Farmville*, there are projects (like *SpaceWarps*) that require lots of clicking. If you prefer puzzles, there are some excellent brainteasers, like the protein-folding game *FoldIt*. In *GalaxyZoo*, there are even hidden object-type setups that require you to examine images to find galaxies. Many of these projects also have badges and titles to unlock as you progress through your mission. Others provide leaderboards and bragging rights. Some even provide credit in academic papers.

So . . . if you're looking for downtime and relaxation but don't want to succumb to *Candy Crush Saga* again, citizen science is for you.

You can find meaning and purpose

Life is a struggle. For the vast majority of people on Earth, just surviving is a struggle.

For those of us lucky enough to have either attained or been born into better circumstances, just making ends meet can be a trial. Beyond that, we strive to self-actualize, and to add meaning and depth to our existence.

There are many different paths to meaning, of course. Some people find what they need in religion; others find it in raising a family. Still others discern the path to meaning through a combination of endeavors.

Is citizen science the path to enlightenment? Probably not. But it does represent at least one method to add depth to what you do. When you participate in a citizen science project, you're contributing in two ways. First, you're helping to increase the sum of knowledge and understanding we have of our universe and our place in it. Second, you're contributing directly to the solution of one of the many problems humanity has.

And wouldn't you know it, having a sense of purpose is good for you, too.[38]

You can make a difference regardless of your skill level or education

Neil deGrasse Tyson, an astrophysicist, currently has 1.3 *million* Twitter fans. This is not as many as, say, Snooki, but still, it's pretty amazing. As of this writing, the Facebook page for ScienceAlert has 3.7 million fans and is growing at a rate of tens of thousands of fans per month. *Star Trek*, and all its reincarnations and reboots, has tens of millions of fans in the United States and around the world. (You may even be one of them.) For that matter, more than 100,000 people have applied for a one-way ticket— essentially, a suicide mission—to Mars, and they paid money just to apply.[39]

An exploration of the motivations involved in science and sci-fi fandom is beyond the scope of this book,[40] but clearly, many people love both science and science fiction. It's also clear that not all those fans are working scientists. The entire US science and engineering workforce was just 4.9 percent[41] of the population in 2010, or around 15 million[42] people.

Citizen science channels all that enthusiasm for science— and the vision of the future it is often thought to represent—into more than just sci-fi conventions, box office receipts, and photo sharing on Facebook. Don't get me wrong: those things are fun and have tremendous value. But citizen science gives ordinary people, from high school students to retirees, a chance to make a genuine contribution to a field they love. The projects are set up to be easy to do, include training where necessary, and have data checking and redundancies built in so you don't have to worry about doing something wrong.

And whether you're sifting through pictures to identify craters on the Moon or listening to whale song, you're doing the same kind of work as Darwin, Mendel, and Barber.

You can fit it into any schedule

Thanks to the proliferation of citizen science projects in the last five years, you can be involved as little, or as much, as you want. If you've got money, but very little time, you can fund individual research studies directly. If you've got the time, you can do everything from exploring Mars (from the comfort of your computer desk) to going out birding every weekend.

In Part II, I have categorized citizen science activities in levels ranging from the easiest to the most complex and interesting, so you can dive in wherever it feels comfortable.

You can pick your favorite topic

With a huge range of topics to choose from, you're bound to find something of interest. If you're a closet math nerd, then installing a program that looks for prime numbers might be your thing. If you love gardening, keeping track of the critters in your patch of the globe may be just the excuse you need to get outside more often. History, biology, physics, paleontology . . . it's all available.

You can learn . . . and protect yourself

There's a proverb that goes something like: "Tell me and I'll forget; show me and I may remember; involve me and

I'll understand." That's especially true when it comes to citizen science.

It's one thing to read about drug development in a magazine article; it's quite another to improve drug-targeting research by actually examining a protein model on a computer screen. When you get involved in a project, you'll get a much better understanding of the topic and how it connects with the wider world.

Achieving this understanding will become increasingly important as we move through the 21st century. We live in an age driven by technology. To make wise decisions, we must understand the world in which we live. Consider genetically modified food: it's an incredibly important issue, and the decisions we make about it have far-reaching implications now and for generations to come. But it's a debate clouded by vested interests, politics, money, and propaganda . . . on both sides. Citizen science allows you to understand the basics of issues like this one, so you can sort through complex and opposing ideas.

Tyson is quoted as saying, "If you're scientifically literate, the world looks very different to you. And that understanding empowers you."

And it's now OK to admit it

There was a brief period between the Moon landing and pre-Apple fanboys when a love of science and technology wasn't very cool. But a funny thing happened on the way to the 21st century: Internet "culture," insofar as there is one, has made science and technology—and by extension, geeks, and nerds—socially acceptable again. Now, people feel more comfortable about expressing an interest in something that was once not all that hip. So if you've been worried about what people might think about this sort of thing, worry no longer. It's cool.

Are you ready to get your geek on? Let's go!

PART II — THE MANY WAYS YOU CAN DO CITIZEN SCIENCE

In this section, I'll list some of the projects you can get involved with right now. New projects are being set up every day, so to stay up to date, or to learn more about projects I've mentioned but don't detail explicitly below, visit www.CitizenScienceCenter.com, where you can sign up for the free mailing list or RSS.

LEVEL 1 — DONATE

Difficulty: Super easy

Requirements: Some spare cash

One of the easiest ways to get involved in citizen science is simply to donate. You might donate to a particular citizen science project you are interested in but perhaps don't have the time for. Most of these projects have a donation button or a donation link somewhere on their site, so I won't list them individually here.

You can also check out one of several crowdfunding sites available and fund a specific research study being conducted by full-time scientists.

If you're not familiar with the concept, "crowdfunding" is a fundraising model used by artists, musicians, writers, and increasingly, scientists, to appeal directly to the public for financial contributions. The idea has its advantages and its disadvantages. On the plus side, scientists can tap into a new source of money and worry less about politics—both national and office—getting in the way. On the down side, it's not yet widely known that the public can fund research directly, so many studies are going begging. Perhaps you can change that.

Incidentally, some crowdfunded projects offer incentives and bonuses, depending on how much you give. These can include anything from exclusive access to videos or e-books, to a say in how the project moves forward, to downloadable goodies, and so on.

Kickstarter

The best-known crowdfunding site, Kickstarter, is mainly a general interest site, but it hosts plenty of science-based projects if you know where to look. Previous projects include "ARKYD: A Space Telescope for Everyone"; "SkyCube: The First Satellite Launched by You!"; "oneTesla: A DIY Singing Tesla Coil"; and "The Smart Citizen Kit: Crowdsourced Environmental Monitoring."

http://www.kickstarter.com/discover/tags/science

Indiegogo

Indiegogo is probably the second-best-known crowdfunding source. Until recently, it was one of the few platforms available outside the United States. Most of the science projects currently listed here are education related (building science apps for kids, for example) or arts related (documentaries about scientific topics), but check back frequently, as projects change all the time.

http://www.indiegogo.com/projects?utf8=%E2%9C%93&filter_title=science&search_submit=Search

Experiment.com (formerly Microryza)

Experiment.com is all about science, and that's evident in both the project names (for example, "Can we push the boundaries of space propulsion?" and "Can intervention improve abilities of children with low motor skills?") and the categories: biology, physics, ecology, chemistry, and so on. Experiment.com focuses on the direct funding of research studies. When you fund research here, you receive online access to the results, with your name in the credits.

https://www.experiment.com

SciFund Challenge

One of the SciFund Challenge group's three goals is to run crowdfunding drives. This group tends to focus on rallying interest around specific, limited-time campaigns. The most recent challenge was "Save Snapshot Serengeti," which was a major success. To jump onto the next campaign, follow the blog or keep an eye on the Twitter hashtag.

http://scifundchallenge.org/blog/

https://twitter.com/search?q=%23scifund

LEVEL 2 — SET AND FORGET

Difficulty: Easy

Requirements: A relatively recent computer[43] and a decent Internet connection

Another really easy way to participate in citizen science is to get involved with something called "distributed computing," which has also been called "grid computing," "volunteer computing," and "networked computing."

In the past, researchers could only tackle fiendishly big or complicated problems with supercomputers capable of doing lots of calculations simultaneously—like the Tianhe-2 in China,[44] which clocks in at 54.9 petaflops, or 54.9 quadrillion calculations per second. These computers are expensive to build—the Tianhe-2 cost about $290 million—and everyone with heavy-duty computational needs wants time on them, so they can be hard to book.

As a solution, computer scientists can now take a really big or complicated problem, divide it into thousands of smaller problems, and then send out those smaller problems via the Internet to thousands of ordinary computers. Most of us have more computing power in our desktop or laptop computers than we actually use, and most of our computers sit idle a lot of the time. By installing a special program on your computer, you can receive one slice of a problem and "donate" your spare computer calculation cycles to solving it.

You can choose from dozens of programs. In this book, I'll concentrate on those that make use of a platform called the Berkeley Open Infrastructure for Network Computing, or BOINC. The list below is not comprehensive, but it will give you an idea of the range of projects you can sign up for.

BOINC

To donate your computer power to the projects listed below, you need to have a piece of software installed on your computer to receive, process, and send problems. You can get that program from the main BOINC site.

http://boinc.berkeley.edu/download.php

Asteroids@home

We have catalogued hundreds of thousands of asteroids, but we know very little about them. Scientists would like to know more about their size, shape, rotation speed, and axis of rotation. In this project, your computer would analyze photometric data from sky surveys to provide a better understanding of asteroids.

http://asteroidsathome.net/boinc/

CAS@home

Here, CAS stands for the Chinese Academy of Sciences, and this name has been a catchall for several projects the academy has run. Previous projects have looked at protein structure prediction and simulating the flow of fluids and the motion of solids on the nanoscale. Upcoming projects include an application for cancer gene research and another application for simulating particle collisions at the Beijing Electron Positron Collider.

http://casathome.ihep.ac.cn/index.php

Climateprediction.net

Climate change is difficult to predict. Our current models all suggest major changes will occur in the next 100 years, but the forecasts vary wildly. This poses challenges to those making policy decisions, or even to getting people to agree that action is necessary. Climateprediction.net aims to get a better understanding of how climate change models work by testing them thousands of times with different parameters and in different scenarios. There are currently four different projects available here.

http://www.climateprediction.net/

Constellation

This platform for doing research in aerospace science and engineering uses your computer to run simulations. Projects in progress right include a trajectory optimization for launchers, satellites, and probes; an analysis of the Moon's near-surface exosphere; and an analysis of exploration rovers' dynamic systems.

http://aerospaceresearch.net/constellation/

Cosmology@Home

Does dark matter exist? Is string theory the right model for describing our universe? Cosmology@Home does research to identify the models that work best with the data we have from particle physics and astronomy.

http://www.cosmologyathome.org/

Einstein@Home

How would you like to search for neutron stars? This project searches for weak astrophysical signals from spinning neutron stars. It also looks for gravitational-wave emissions, which were predicted by Albert Einstein but have never been observed directly.

http://einstein.phys.uwm.edu/

Enigma@Home

In an attempt to prevent their country's enemies from understanding intercepted messages during World War II, the Germans used "Enigma machines" to encrypt communications. Work by the Poles in 1932, and later by a dedicated staff at Bletchley Park in England,[45] cracked the code, and the intelligence gathered helped win the war. Some messages remain unbroken, however, and this project is designed to decrypt them.

http://www.enigmaathome.net/

FightMalaria@Home

This research project does docking simulations on malaria proteins to try to develop the drugs that most effectively treat this disease.

http://boinc.ucd.ie/fmah/

GPUGRID.net

This is another project that performs molecular simulations to understand the function of proteins in health and disease. Specifically, it examines HIV, cancer, and neural disorders.

http://www.gpugrid.net/

MindModeling@Home (Beta)

If brains are your thing, you'll like this: this beta project focuses on using computational cognitive process modeling to explain the processes that enable human performance and learning.

http://mindmodeling.org/

Quake-Catcher Network Sensor Monitoring

There are two parts to this project, and you can participate in either part—or both. In the first part, volunteers in quake-prone areas hook up movement sensors to their computers. If your home experiences any rumbles, it will transmit that data back to the QC Network. In the second part, volunteers with BOINC installed and QCN enabled will process those measurements and provide information to researchers about quake distribution and size.

quakecatcher.net

Rosetta@Home

This project aims to figure out the three-dimensional shapes of proteins to find cures for major diseases such as HIV, malaria, cancer, and Alzheimer's. For example, researchers believe abnormal protein folding, in which proteins form structures called *amyloids* instead of folding normally, causes Alzheimer's. Scientists want to predict which parts of proteins are likely to form amyloids so they can ultimately block these formations.

http://boinc.bakerlab.org/rosetta

SETI@Home

This is perhaps the most "famous" BOINC project. SETI is short for "search for extraterrestrial intelligence" (yes, aliens). With SETI software installed, you'll be scanning radio telescope data for signals from the heavens.

http://setiathome.berkeley.edu/

Zooniverse

Last, but definitely not least, Zooniverse is the place to go if you want access to a range of web-based citizen science projects. Over the years, its organizers have perfected how to set up, manage, and complete projects, and so a lot of scientists are using the platform. Bonus: you only need to set up a login here once, and it keeps stats on all of your activity. There are usually more than fifty different projects to choose from!

https://www.zooniverse.org/

LEVEL 3 — WEB-BASED CITIZEN SCIENCE

Difficulty: Still easy

Requirements: An up-to-date computer, a fast Internet connection, and some spare time

While donating and downloading are great ways to get involved in citizen science, they aren't quite as much fun as actually *doing* citizen science. So in this section, we're going to look at projects in which you can be directly involved and thereby contribute to research.

I've called this level "web-based" because all the projects described below have a website you log into and work with. In most cases, you will either be looking at images or videos, or listening to sounds, and classifying what you're seeing or hearing. In a few of the listings, you'll be doing something closer to playing a game (so yes, you can tell your mother that all that time spent with the console was actually *training*). There are an even wider range of topics to choose from at this level, because the projects aren't restricted to what a computer can do.

EteRNA

How would you like to "play" with the stuff of life? EteRNA is a puzzle-solving game based on ribonucleic acid (RNA) structures. RNA is a large molecule that assists with coding, decoding, regulating, and expressing genes. Scientists are still figuring out how RNA works.

As a player, you're presented with a target shape and challenged to fold an RNA strand into that shape. You do this by changing the order and placement of the four RNA building blocks—colored dots representing adenine, cytosine, guanosine, and uracil—at various positions. This will change the way the RNA twists, turns, and forms a shape. The game is a visual treat with high-quality graphics; you also level up and receive points as you progress through the challenges.

By solving the puzzles presented, EteRNA players help create a large-scale library of synthetic RNA designs, which will be used to create RNA-based switches and nanomachines that will seek and control things like disease-causing viruses.

http://eterna.cmu.edu/web/

Flu Near You

Nobody likes to be sick, but if you join Flu Near You, you can at least use your sniffles for science.

Flu Near You is an illness-tracking website designed to give researchers and health-care professionals a better idea of influenza trends in real time. To participate, you register at the site, provide a bit of basic information (things like age, gender, postal/zip code), and then agree to fill out weekly surveys. You can also visit the site regularly to see if people in your area have started to report symptoms. The site also provides information on flu shot availability.

https://flunearyou.org/

Orcasound.net

Whales are fascinating creatures and their social structures are still poorly understood. Researchers with the Salish Sea Hydrophone Network want you to listen for whale song and tell them when you hear it.

A hydrophone is an underwater microphone, and there's a network of them set up near Seattle and Vancouver. To participate, you simply click on a live-feed link and let the sounds of the sea wash over you. If you should hear some whale song, you send a note to organizers via email or log the sound in a shared Google doc.

A sound tutor will even help you identify particular whale "pods" (groups) by their song. This would be a great project for you if you like quiet background noise when you're working at the computer.

http://www.orcasound.net/

Happy Match

You may have played a classic game that goes by the name of *memory* or *concentration*. Happy Match, located on the Citizen Sort website, has applied that concept to a citizen science project.

In this case, players participate in the research by classifying photos of animal, plant, and insect species.

There are different versions of Happy Match for moths, sharks, and rays. The game is set up so you can score points at the end of each game and compete with friends. The game also serves a second purpose: to learn more about what motivates and retains citizen scientists.

http://citizensort.org/web.php/happymatch

TestMyBrain

Admit it: you have taken at least one "personality test" in your life, right? Perhaps it was one that answers the question, "Which celebrity are you most like?" Or maybe it was a more serious survey to determine what kind of learning style suits you best.

Answering questions to learn more about yourself can have a lot of value. And in the case of TestMyBrain, you'll also be helping psychologists learn more about how the brain works.

There are currently five tests available, and the total time commitment is roughly one hour. No personal information is collected.

http://www.testmybrain.org/index.php

The Baby Laughter Project

Who doesn't love the sound of babies laughing? While the rest of us find baby giggles good for a smile, researchers at the Birkbeck Babylab (yes, there is such a thing!) believe studying infants cracking up can help us learn about human brain development. If they can figure out what babies find funny, they'll have insight into how babies learn about the world. They'll also learn more about that strangely human characteristic, a sense of humor.

You can participate in three ways. If you're the parent or guardian of a child less than two and a half years old, you can fill out a short, anonymous, and confidential survey about your child. If you don't have a child in the indicated age range but occasionally interact with one, you can submit a "field report" about a particular incident that made a baby laugh. And if you have a video of a baby laughing, the project would love to hear about it.

http://babylaughter.net/

The Milky Way Project

How is a star born? On Earth, it might begin with *American Idol*. But out in space, it could begin with a bubble.

Massive clouds of dust and gas collapse to form stars. We often see "bubbles"—which look much like you'd expect—in areas of star formation, but we don't understand how they work in a star's life cycle.

The Milky Way Project wants you to find and identify bubbles while looking at images from the Galactic Legacy Infrared Mid-Plane Survey Extraordinaire (GLIMPSE) and the Multiband Imaging Photometer for Spitzer Galactic Plane Survey (MIPSGAL). By learning more about where bubbles appear, we can learn more about their role in galactic space.

http://www.milkywayproject.org/

Seafloor Explorer

For some time now, scientists have been using HabCam, a habitat camera mapping system, to take images of the sea floor. In about a year, they have amassed a collection of more than 30 million high-resolution images, or more than 30 terabytes of data.

Now they need help in classifying the images. Specifically, they want to know what kind of ground cover is in the image (sand? cobble? shells?) and what creatures are visible in the image (scallops? sea stars? fish?). Using a simple point-and-click interface, you'll even be able to provide measurements of the species you see.

This project is designed to build a robust and searchable database to answer important ecological questions that will help inform both future studies and public policy.

CHANDRA CLARKE

http://www.seafloorexplorer.org

Worm Watch Lab

If squirmy things don't squick you out, then watching films about worms might be the project for you.

In this project, scientists are trying to identify connections between genes and behavior. The study is based on the idea that if a gene truly influences an observable behavior, then we should be able to determine this by creating mutations that break that gene, and then watching to see if and how behavior changes.

In Worm Watch, scientists focus on how mutations affect the egg-laying behavior of the nematode worm *C. elegans.* These worms are teeny, have very short life spans, and are easy to keep and handle. Project directors have used automatic tracking microscopes to create videos of the worms (talk about living under the microscope!). Your job is to watch the 30-second clips and click a button if you see a worm lay an egg.

http://www.wormwatchlab.org

LEVEL 4 — APP-BASED CITIZEN SCIENCE

Mobile devices such as netbooks, tablets, smartphones, and smartwatches, and cheap sensors and transmitters like RFID tags are changing the way we think about computing and the Internet. Concurrent with the hype about Web 2.0 and social media is the concept of an "Internet of things"[46]—that is, a network of billions of devices constantly transmitting and receiving data, either automatically or with human intervention.

Even as the citizen science movement takes off, it is also evolving to keep up with the times. Many new projects are now "app-based," meaning you can install an app on your tablet or smartphone to contribute to citizen science.

With web-based citizen science, you're most often asked to help *process* data. With app-based citizen science, you're usually *collecting* data. The difference in approaches makes sense, considering the channels involved.

In the listings below, I'll give you the title of the app and note whether it's available for iOS or Android devices (or both!), so you can search for it and install it via iTunes or Google Play, respectively.

CreekWatch (iOS)

Creeks are an important part of your local ecosystem: they provide water for both plants and animals. But there are too many creeks for water control boards to monitor effectively. You can use this app to take photos of your local waterway(s) and answer a few simple questions about water and (sadly) trash levels.

Firefly Flash Counter (iOS)

Fireflies are an integral part of childhood summers for many people, but are populations declining? This app helps you count and report the number of flashes you see at a given location.

The Great Brain Experiment (iOS, Android)

This project is designed to test your memory, your impulsivity, your attention, and your decision-making abilities by getting you to play games that are actually neuroscience experiments. Researchers get to conduct tests on a much bigger scale, and you'll find out whether you're faster or more attentive than other people, and whether you can remember more or make better choices.

iNaturalist (iOS, Android)

Part social network for naturalists, part citizen science project, this app allows you to record your observations of different plants and animals and contribute them to a growing online database. The founders' vision is that, "if enough people recorded their observations, it would be like a living record of life on Earth that scientists and land managers could use to monitor changes in biodiversity, and that *anyone* could use to learn more about nature."

Instant Wild (iOS)

In this app, images of wild animals are sent to you automatically from webcams placed in remote locations. You identify the animals you see by matching them with images in a field guide. The goal here is to save conservationists thousands of hours by sorting images, which enables them to analyze the data much faster. This is particularly important for making effective conservation decisions.

Marine Debris Tracker (iOS, Android)

This one is a bit like Foursquare.com for trash. When you're near a body of water, you "check in" and report whether or not you see trash and, if so, what kinds. If it appears safe to do so (that is, the debris is not too large to handle, a dead animal, or otherwise potentially hazardous), you can also conduct a cleanup and dispose of or recycle the material properly.

Sealife Tracker (iOS)

If you're a diver and you live in the UK (or just need an excuse to visit there), this app is for you.

Researchers with the British Sub-Aqua Club have developed an app that will enable divers to help monitor the spread of both invasive and climate-change indicator species in British seas. Before a dive, you check out the app to see what species are of concern. After a dive, you can register your observations.

Splatter Spotter (iOS)

It's a disgusting name for an app, but it's a disgusting problem: roadkill. Estimates put the number of vertebrate animals (those with backbones) killed by vehicles at 1 million per day in the United States alone.

This application wants you to mark the location and, if you can tell, the type of animal that was killed. The goal is to see if we can learn more about what species are being affected and how drastically. Do *not* use this app while driving. Not only is this illegal in most jurisdictions, but you do not want to cause—or become—roadkill yourself. Either have a passenger log the data, find a safe place to pull over, or fill in the details when you have finished your journey.

LEVEL 5 — GET OUTDOORS FOR CITIZEN SCIENCE

Technology has helped citizen science projects proliferate, but it's not all about screens and keyboards; plenty of projects let you geek out in the fresh air. This is good for you, and for science. Check out the projects below.

Butterflies and Moths of North America (BAMONA)

There are few things as beautiful or delicate to look at than a butterfly or moth. The BAMONA project wants to know more about the *lepidoptera* in your neighborhood, and to get real-time data about species distributions and populations. For this project, BAMONA invites you to submit clear, high-quality photographs to their website at any time. If you're a deft hand at identification, you can also volunteer to review and classify photos. (You could also combine your work in this project with Project Bumble Bee, described below.)

http://www.butterfliesandmoths.org/

Camel Cricket Survey

A part of the "Your Wild Life" series of projects, the Camel Cricket Survey wants to know about native and invasive species of the camel cricket. Also known as "sprickets" or "criders" (shortened from "spider crickets"), these insects can be found on every continent in cool, damp, and dark places.

Good places to look include your basement and garage, nearby wells, under leaves and stones, and in stumps and hollow trees. Snap a picture and report your findings at any time.

http://www.yourwildlife.org/projects/camel-crickets/

Project Bumble Bee

You have no doubt heard about the honeybee crisis: colonies are dying *en masse* for reasons that still aren't quite clear. Bumble bee populations are also dwindling, and they, too, play an important role in pollinating both wild plants and domestic crops.

There are two ways to participate in this project. One is to take direct action to conserve your local population by planting flowers bumble bees like. (Bonus one: you get a lovely flower garden you can enjoy. Bonus two: butterflies love many of the same plants, so you'll probably help conserve those, too.) The second way to participate is to send in photos of the bees in your area to give scientists on-the-ground information about where bee populations are.

http://www.xerces.org/bumblebees/

Project FeederWatch

One of the saving graces of a long, cold winter is the sight of birds at the feeder. This project wants you to pay a bit more attention to who is visiting your feeder from November to April. When you register for this project, you pay a small fee to receive a kit and a field guide to help you make accurate observations.

FeederWatch data helps scientists track movements of winter bird populations and trends in bird distribution and populations.

http://www.birds.cornell.edu/pfw/Overview/over_index.html

Project Squirrel

We sort of take squirrels for granted; they're everywhere, and they seem to thrive in both urban and nonurban environments. But their ubiquity is precisely the reason they need monitoring: watching them can tell us how our environment is doing.

Participating is easy—just keep an eye on your local squirrel population and record your observations regularly at the Project Squirrel website. You can contribute photos and stories, and read what others have posted as well.

http://www.projectsquirrel.org/index.shtml

SharkTrust: The Great Eggcase Hunt Project

Judging by the popularity of Shark Week, a TV event in the United States, sharks continue to frighten and fascinate us. As scary as they seem, sharks are a vital part of the ocean ecosystem, and much work remains to be done to conserve them.

If you live in the UK, you can help by hunting for eggcases the next time you visit the seaside. An eggcase, or "mermaid's purse," is the tough envelope protecting a shark embryo; they also protect skate or ray embryos. Eggcases frequently wash up on the shore, and reporting a find will help researchers know how shark populations are doing.

In addition to the Great Eggcase Hunt, there are several other shark-related projects on the Shark Trust site.

http://www.sharktrust.org/en/GEH_the_project

Spiders in Your World

They're not everyone's favorite critter, but spiders play an important role in pest control and the ecosystem. This project wants to know in which climates certain spiders live, and to track spider distribution over time. Among other things, the data will help us learn more about how climate change might be affecting our friends with eight legs.

You can photograph and log your spider sightings through the iNaturalist site, at the link below.

http://www.inaturalist.org/projects/where-is-my-spider

WHERE TO GO FROM HERE

This book has only scratched the surface of the citizen science movement. There are dozens more projects to participate in. Some are very local and focus on specific areas of the world, while others are national or international. And as some projects draw to a close, new ones come online.

To learn about new projects as they emerge, or to learn about projects not covered here, visit www.CitizenScienceCenter.com. There's a free newsletter that goes out once or twice per week, as well as an RSS feed at www.CitizenScienceCenter.com/feed. If Twitter is your thing, follow @citizenscience_ (note the underscore at the end).

 ** Finally, if you liked this book, it would mean the world to me if you left a review for it at Amazon. Positive reviews help sales, and my goal is to let as many people know about citizen science as possible! **

BONUS RESOURCES!

The citizen science projects I have covered in this book and on my website only scratch the surface of what is available to the science enthusiast. In this bonus section, I provide you with dozens of other ways to get in touch with your inner nerd.

FREE SCIENCE EDUCATION OPPORTUNITIES

CosmoLearning: A science education aggregation site that lists more than 515 courses from around the globe in every conceivable topic.
http://www.cosmolearning.com/courses/

Coursera: A consortium of US universities offering lecture-based courses in several science-related areas via online videos.
https://www.coursera.org/

EdX: MIT and Harvard, among others, have joined forces to offer free courses to a global audience.
http://www.edxonline.org/

FutureLearn: This site offers courses from a group of UK and international universities.
https://www.futurelearn.com/

HippoCampus: A variety of high school and college courses from the US National Repository of Courses Online.
http://www.hippocampus.org/

iTunes University: Carry your university-level education in your pocket. This link directs you to the web page, but you can also access the app by searching for iTunes University in the Apple App Store on your favorite device.
http://www.apple.com/education/itunes-u/

Johns Hopkins: This open courseware initiative by Johns Hopkins University focuses on medical science topics.
http://ocw.jhsph.edu/index.cfm/go/find.browse#topics

Khan Academy: With more than 3,200 videos, you could spend months on this site alone. While there is not much K-8 material, there is a significant amount at the high school and college level.
http://www.khanacademy.org/

Learners TV: This site offers videos, lecture notes, and course descriptions on topics ranging from biology to economics.
http://www.learnerstv.com/

Math.com: Everything from basic math to trigonometry can be found on this site, with a few fun math games thrown in for good measure. There is also a store where you can buy books, CDs, software, etc.
http://www.math.com/

Merlot: Teachers and professors can share their best material at the Merlot repository. You can grab some learning goodies here, too.
http://www.merlot.org/merlot/materials.htm?category=2
605&sort.property=overallRating

MIT: More than 2,100 undergraduate- and graduate-level courses are available on this site, which focuses heavily on science-related material.
http://ocw.mit.edu/courses/

Open Learn: This site offers eight pages of free science courses from the Open University in the UK.
http://openlearn.open.ac.uk/

Open Learning Initiative: Carnegie Mellon University offers college-level courses in more than twelve areas,

ranging from argument diagramming to statistical reasoning.
http://oli.cmu.edu/learn-with-oli/see-our-free-open-courses/

Science media: Personally, I prefer books to videos. You can find top-notch reviews of science books and films on this website from the American Association for the Advancement of Science (AAAS).
http://www.sbfonline.com/Pages/welcomesplash.aspx

Science NetLinks: This site offers the children in your life free K-12 science education from the AAAS.
http://sciencenetlinks.com/

SciTable: Nature brings you material on cell biology and genetics.
http://www.nature.com/scitable

Stanford Engineering: The Stanford School of Engineering allows free access to some of its most popular courses online.
http://see.stanford.edu/see/courses.aspx

TED: This site offers technology, education, and design talks from some of the most interesting people in the world.
http://www.ted.com/
PS – Check out my TEDx talk on citizen science:
https://www.youtube.com/watch?v=U7XOcB6_TWw

Udacity: If you're looking for a different type of learning experience, this site offers courses that use a problem-solving approach rather than a lecture format. The site's content is focused on computer science.
http://www.udacity.com/

University of California at Berkeley: This site includes everything from anthropology to robotics.
http://www.virtualprofessors.com/directory/university/berkeley

University of California at Irvine: On this site, you can learn about the physical sciences, engineering sciences, and computer sciences.
http://ocw.uci.edu/courses/?cat=9

University of Michigan: A good list of material focusing on health sciences at the college level.
https://open.umich.edu/education

University of the People: A tuition-free university that is designed to democratize education. There are fees to apply and take exams, but these max out at $100. At the time of writing, they were offering an undergraduate Computer Science degree.
http://www.uopeople.org/

Yale: This Ivy League school provides a list of free videos, suggested readings, and practice problems.
http://oyc.yale.edu/courses

YouTube Edu: Believe it or not, there is more to YouTube than cute kittens. Have a look at the Edu channel for a wide variety of lectures.
http://www.youtube.com/education

SCIENCE MAGAZINES TO KEEP CURRENT

Air & Space: This magazine covers the history, culture, and technology of flight.
http://www.airspacemag.com/

America Archeology: Providing information on the latest discoveries, current research, and other news, this site focuses on archeology in the Americas.
http://www.americanarchaeology.com/aamagazine.html

American Scientist: A magazine that looks at research projects, researchers, techniques, and technology related to scientific developments.
http://www.americanscientist.org/

Astronomy: Telescopes, comets, stars, planets, the sun and moon, constellations, and observation techniques provide the main content for this magazine.
http://www.astronomy.com/

Astronomy Now: A magazine for astronomers and space enthusiasts, offering news, information, reviews, features, and new products.
http://www.astronomynow.com/

Audubon Magazine: With a particular focus on birds and their habitats, this magazine is designed to help readers appreciate and preserve the natural world.
http://www.audubonmagazine.org/

Aviation Week & Space Technology: This magazine reports on technical, legislative, scientific, operational, and

financial developments in the military, commercial, and space aviation markets.
http://www.aviationweek.com/

Connected World: Focuses on how we use technology for both work and life.
http://connectedworldmag.com/

Current World Archaeology: This magazine reports on digs and discoveries from around the world.
http://www.archaeology.co.uk/

Elektor: Each issue of *Elektor* magazine features construction projects and discusses electronics for professionals, enthusiasts, and students.
http://www.elektor.com/

Food Technology: This magazine covers all aspects of food, including biology, chemistry, nutrition, engineering, production, microbiology, packaging, quality assurance, regulations, research and development, and even consumerism.
http://www.ift.org/food-technology.aspx

Green Source: The design and construction of environmentally responsible buildings make up the content of this publication.
http://greensource.construction.com/

Issues in Science & Technology: A forum for the exchange of ideas on policy issues related to science, technology, and health.
http://www.issues.org/

MAKE Magazine: This magazine discusses inventing and invention and has lots of DIY projects.
http://makezine.com/

Mineralogical Record: This is a publication for serious mineral collectors.
http://www.minrec.org/

National Geographic: A general interest magazine that focuses on natural history, geography, and wildlife.
http://www.nationalgeographic.com/

Natural History Magazine: This publication focuses on nature, science, and culture.
http://www.naturalhistorymag.com/

Nature: An international journal publishing peer-reviewed research in all fields of science and technology.
http://www.nature.com/

New Scientist: A generalist publication that covers a wide range of scientific topics, including botany, physics, evolution, nuclear power, mathematics, and the environment.
http://www.newscientist.com/

Nuts & Volts: This magazine is intended for the hobbyist, design engineer, technician, or experimenter. Topics in this publication include robotics, circuit design, lasers, computer control, home automation, microcontrollers, new technology, and DIY projects.
http://www.nutsvolts.com/

Physics Today: This magazine reports on the latest advances in physics and related sciences.
http://www.physicstoday.org/

Popular Mechanics: Gives you the latest news and trends in automobiles, home improvement, tools, electronics, health, science, and technology.
http://www.popularmechanics.com/

Popular Science: A magazine that covers the latest science and technology developments in computers and electronics, aviation, space, automobiles, medicine, energy, and consumer electronics.
http://www.popsci.com/

Science Illustrated: From paleontology to space exploration, this magazine publishes articles related to a variety of cutting-edge scientific issues.
http://www.sciencenews.org/

Science News: A bi-weekly news magazine that covers the most important research findings in all fields of science and medicine.
http://www.sciencenews.org/

Scientific American: A publication that produces articles and commentary on the latest science news.
http://www.scientificamerican.com/

Sea Technology: This magazine offers information on marine business, science, and engineering for commercial and military applications.
http://www.sea-technology.com/

SEED Magazine: Covers science and its effects on society.
http://seedmagazine.com/

SERVO Magazine: Published for robotic experimenters, each issue of SERVO contains feature articles, interviews,

tutorials, projects, and sources for parts.
http://www.servomagazine.com/

Sky & Telescope: A resource magazine for amateur astronomers.
http://www.servomagazine.com/

Smithsonian Magazine: This periodical covers topics that include nature, history, science, and the arts.
http://www.smithsonianmag.com/

T3: Gadgets and technology are the topics of interest in this magazine.
http://www.t3.com/

Technology Review: Featuring information and analysis on emerging technologies, this magazine focuses on trend-setting innovations and their economic, commercial, social, and political impact.
http://www.technologyreview.com/

Wired Magazine: Covers science, technology, emerging trends, and entrepreneurialism.
http://www.wired.com/

GET TINKERING – OPEN SOURCE SOFTWARE AND HARDWARE SITES

Arduino: Arduino is an open source electronics platform with easy-to-use hardware and software.
http://arduino.cc/

Arduinome: A MIDI-controller device for the Arduino physical computing platform, Arduinome is released under an open source, non-commercial-use-only license.
http://flipmu.com/work/arduinome/

BeagleBoard: A low-cost, single-board computer with an ARM processor, 128 Mb memory, SVideo, USB, and some PC peripherals.
http://www.beagleboard.org/

Bowden's Hobby Circuits: This website details over 100 circuit diagrams for commonly used circuits, as well as links to related sites, commercial kits and projects, newsgroups, and educational areas.
http://www.bowdenshobbycircuits.info/

Carambola: A tiny (35x45 mm), cheap, open source module that adds wireless and wired networking capabilities to any device.
http://8devices.com/carambola

Chispito Wind Generator: A guide outlining how to build your own Chispito wind generator.
http://www.velacreations.com/energy/electrical-sources/wind-power/item/77-chispito-how-to.html

ClockTamer: A low-cost, small-size, configurable clock generator with a GPS sync option.
http://code.google.com/p/clock-tamer/

ColorHug: An open source display colorimeter that allows you to calibrate your screen for accurate color matching.
http://www.hughski.com/

CubeSpawn: This one bills itself as an open source "flexible manufacturing system" or FMS. Modular design.
http://www.cubespawn.com/

CXI: Circuit Exchange International is a database of ready-designed circuits.
http://www.zen22142.zen.co.uk/

Discovercircuits.com: A collection of more than 11,000 schematics cross-referenced into more than 500 categories.
http://www.discovercircuits.com/list.htm

DSO Nano: An open source pocket-sized digital oscilloscope based on the STM32 microcontroller IC.
http://code.google.com/p/dsonano/

Educational colorimeter kit: An affordable colorimeter kit for educators, students, and citizen scientists.
http://www.iorodeo.com/colorimeter

Electronics and Radio Today: This site offers free information, articles, and projects for the radio and/or electronics hobbyist. http://www.electronics-radio.com/

Electronics Lab: This website offers visitors electronic circuit descriptions, diagrams, electronics articles, links,

downloads, and a community message board.
http://www.electronics-lab.com/

Electronics Project Design: Practical electronics project schematics, parts lists, component descriptions, product testing, and other references for the electronics hobbyist and electronics designer are offered on the Electronics Project Design website.
http://www.electronics-project-design.com/

Elphel: This open-hardware and open source camera can be customized for many different applications.
http://www3.elphel.com/index.php

E-Music DIY Archive: An archive that provides schematics, manuals, references, and information related to electronic music.
http://www.emusic-diy.org/

Ethernut: An open source hardware and software project for use as an embedded ethernet system.
http://www.ethernut.de/

FC's Electronic Circuits: A free circuit archive.
http://www.solorb.com/elect/

FLASH-PLAICE: An open source in-circuit development tool that combines a FLASH programmer, emulator, and high-speed multi-channel logic analyzer on one board.
http://flash-plaice.wikispaces.com/

Freeduino: A repository of files that allow you to make circuit boards that are compatible with Arduino hardware.
http://www.freeduino.org/

CHANDRA CLARKE

FreePCB: With FreePCB, you can design your circuit boards on your PC first. This free open source PCB editor for Microsoft Windows was released under the GNU General Public License and is easy to learn and use.
http://www.freepcb.com/

Fritzing: This is an open source initiative designed to get people to think of electronics as a creative medium. There is also a starter kit to teach basic electronics in a fun and easy way.
http://www.fritzing.org/

Hack a Day: A site for technology enthusiasts who love to tinker.
http://www.hackaday.com/

HackMe Electronics: An open source electronics site with a focus on audio electronics.
http://hackmeopen.com/

Homo ludens electronicus: This site offers free electronics projects, including schematics.
http://ludens.cl/Electron/Electron.html

Instructables: With Instructables, you can share what you make with the world and find all kinds of instruction sets for just about any project you can imagine.
http://www.instructables.com/

LAOS Laser: An open source alternative for laser cutting.
http://www.laoslaser.org/

Lasersaur: An open source laser cutter for makers, artists, and scientists.
http://labs.nortd.com/lasersaur/

Laurier's Handy Dandy Little Circuits: On his site, Laurier Gendron offers free circuits and projects, complete with instructions and schematics.
http://members.shaw.ca/roma/

LeanXcam: An open source–based smart camera that is used for industrial applications in the field of machine vision.
http://www.systronics.ch/eng/vision-systems-leanXcam-starterkit_81433.shtml

Les lampes radio: For you old-school enthusiasts, Claude Paillard constructs his own vacuum tubes by hand. The site is written in French but includes several photos and a 17-minute video showing Paillard constructing a vacuum tube from scratch.
http://paillard.claude.free.fr/

LTspice IV: A free simulation and design software for analog and mixed signals that will allow you to evaluate circuits. This software runs on Windows.
http://www.linear.com/designtools/software/#Ltspice

Light Buckets: You can rent telescope time, so you can do astronomy no matter where you live in the world.
http://www.lightbuckets.com/

Magic Mirror: This tool allows you to make a magic mirror that plays animations based on input from various sensors. It features four characters, each with its own personality.
http://diymagicmirror.com/

microDrum: With microDrum you can make your own open source electronic drum.
http://microdrum.altervista.org/

MikroKopter: Build your own micro-helicopter.
http://www.mikrokopter.de/ucwiki/en/MikroKopter/

Milkymist: An informal organization of people and companies who develop, manufacture, and sell a comprehensive open source solution for the live synthesis of interactive visual effects for VJs.
http://milkymist.org/

MintyTime: This project allows you to build your own binary clock in an Altoids tin. It includes both electronics and a physical component.
http://www.mintytime.com/

Multimachine: This is an all-purpose, open source machine tool that you can build from discarded car and truck parts using commonly available hand tools. It does not require electricity.
http://opensourcemachine.org/

Mutable Instruments: With these kits, you can build your own synthesizers and MIDI processors.
http://mutable-instruments.net/

MyCPU: A homebrew central processing unit that can run as a bitty web server.
http://www.mycpu.eu/

OMFootCtrl: An open source project for designing inexpensive OSC and MIDI foot controllers.
http://omfootctrl.sourceforge.net/

Open Electronics: There are all kinds of very interesting projects here, including how to adapt a 3D printer into a CNC milling machine.
http://www.open-electronics.org/

Open Source Ecology (OSE): The global village construction set (GVCS) is a project that aims to distill modern civilization into 50 different *industrial machines.*
http://opensourceecology.org/

OpenCores: Billing itself as the world's largest open source hardware community, OpenCores develops digital open source hardware through electronic design automation.
http://opencores.org/

Openmoko: A project dedicated to delivering mobile phones that use an open source software stack.
http://wiki.openmoko.org/wiki/Main_Page

openPicus: An Italian company producing hardware for the "Internet of Things." The modules are powered by an open source OS (FreeRTOS), a dedicated TCP/IP software stack, and an embedded web server.
https://code.google.com/p/openpicus/

OpenSPARC: An open source processor project with UltraSPARC T1 and UltraSPARC T2 multicore processor designs contributed by Sun Microsystems.
http://www.opensparc.net/

Osloom: An open source thread-controlled jacquard loom.
http://www.osloom.org/

Pandora: A handheld game console/PC that uses open source software.
http://openpandora.org/

Pinguino: An open source electronics prototyping platform based on PIC (microchip).
http://pinguino.cc/

Ponoko: With this site, you can make and design your own products with 3D printing and laser cutting.
http://www.ponoko.com/

Processing: A programming language, development environment, and online community that has promoted software literacy within the visual arts since 2001.
http://www.processing.org/

Project VGA: A low-budget, open source, VGA-compatible video card.
http://wacco.mveas.com/

Propeller: These microcontrollers, kits, and development boards are programmable in Spin.
http://www.parallax.com/propeller/

RepRap: A self-replicating manufacturing machine that allows you to download the design of the printer and put it together with pieces you buy; the machine can then print the rest of what it needs.
http://www.reprap.org/wiki/Main_Page

RHINO Platform: This acronym stands for Reconfigurable Hardware Interface for computiNg and radiO. It runs on the BORPH OS.
http://www.rhinoplatform.org/

RobotProg: With RobotProg, you can learn programming. Program a virtual robot with a flowchart by first drawing the flowchart and then running the program to watch the robot execute it.
http://www.physicsbox.com/indexrobotprogen.html

Simple Circuit Diagram: On this site, various circuits are organized by category, including audio and music,

oscillators, power supply, radio frequency, security, and test and measurement.
http://www.simplecircuitdiagram.com/

Simputer: A self-contained handheld computer first released in 2002, this open-hardware, Linux-based computer is a low-cost alternative to personal computers.
http://www.simputer.org/

Techlib.com: This site offers free electronics circuits organized by category, including hobby ideas, science, and reference.
http://www.techlib.com/

The Bus Pirate: A universal electronic open-hardware tool designed to simplify the prototyping process.
http://code.google.com/p/the-bus-pirate/

Tinker Forge: A collection of open source hardware modules that can control motors and sense the environment.
http://www.tinkerforge.com/

Touchkit: Touchkit is an open source multi-touch system.
http://labs.nortd.com/touchkit/

TuxPhone: A project to develop an open source (hardware and software) GSM/GPRS cellphone.
http://sourceforge.net/projects/tuxphone/

Ultimaker: An open-hardware 3D printer.
http://www.ultimaker.com/

Wiring: Wiring is an open source programming framework for microcontrollers intended for artistic communities.
http://wiring.org.co/

Xcircuit: An open source circuit-design program for Unix/X11. http://opencircuitdesign.com/xcircuit/

Zoybar: An open R&D lab for academics, hobbyists, and commercial developers to create musical instruments and applications.
http://www.zoybar.net/main/

FUN AND GAMES

Abstruse Goose: A cartoon about math, science, and geek culture.
http://abstrusegoose.com/

Archie McPhee: This site offers kooky science gifts for purchase.
http://www.mcphee.com/shop/

BodyBrowser: An impressive, dynamic map of the human body.
http://www.zygotebody.com/

BrainPop: Animated science and technology content.
http://www.brainpop.com/

Chem4Kids: Chem-4-kids is a good introduction to chemistry. There is also a biology section.
http://www.chem4kids.com/

ChemicalElements.com: This site offers an online interactive periodic table.
http://www.chemicalelements.com/

Chrome Experiments: A site showcasing creative web experiments, a vast majority of which are built using the latest open technologies, including HTML5, Canvas, SVG, and WebGL.
http://www.chromeexperiments.com/

Curiosity: A Discovery Channel site that answers viewers' questions and provides informative articles.
http://curiosity.discovery.com/

Edheads: Free science and math games are offered on this site, including "Simple Machines," "Virtual Knee Surgery," and "Stem Cell Heart Repair."
http://www.edheads.org/

eSkeletons Project: A website devoted to the study of human and primate comparative anatomy.
http://www.eskeletons.org/

Extreme Science: A site dedicated to the biggest and often scariest natural phenomena.
http://www.extremescience.com/

FixYa: A problem-solving service that connects people with experts in a variety of topics.
http://www.fixya.com/

Fun Science Gallery: Have you ever wanted to make your own microscope, telescope, or herbarium? This site allows you to do so.
http://www.funsci.com/

Geek.com: This site delivers information on hardware, software, and gaming, as well as news postings on technology with added discussions.
http://www.geek.com/

GeekSugar: A leading technology guide from Sugar Media Inc.
http://www.geeksugar.com/

Gizmodo: The go-to authority for gadget news and digital culture.
http://gizmodo.com/

HacknMod: One of the biggest hacking and modding communities on the web. Tutorials, guides, and step-by-step video lessons are offered on this site for people to learn to hack and modify common (or retro) game consoles, such as the Xbox 360, Wii, PSP, NES, and Atari.
http://hacknmod.com/

Hacked Gadgets: A how-to site that teaches you how to hack into just about anything.
http://hackedgadgets.com/

HijiNKS ENSUE: A geek pop-culture webcomic that makes fun of the latest news in TV, movies, Sci-Fi, technology, and the Internet.
http://hijinksensue.com/

History of the Universe: Outlines the story of the universe in several chapters.
http://www.historyoftheuniverse.com/

HowStuffWorks: This site offers a vast array of insights into everything from car engines to search engines, cell phones to stem cells, and thousands of subjects in between.
http://www.howstuffworks.com/

Howtoons: This website includes both comic pages and science projects.
http://www.howtoons.com/

Improbable Research: This website covers "research that makes people LAUGH and then THINK." The organization publishes the magazine *Annals of Improbable Research* and administers the infamous Ig Nobel Prizes.
http://www.improbable.com/

Inventive Kids: This site provides information related to several topics, including inventions and women innovators, and offers a variety of activities and games.
http://www.improbable.com/

io9: A site that discusses entertainment, science, and futuristic culture.
http://io9.com/

Kinetic City: A website from the AAAS. Here children can play "kid science detective" games.
http://www.kineticcity.com/

LifeHacker: This site curates tips, tricks, and technology to improve your daily life.
http://lifehacker.com/

MakeUseOf: A booming daily blog that features cool websites, computer tips, and downloads in an effort to make you more productive.
http://www.makeuseof.com/

Moon Phases: This site highlights information related to the moon, including current moon news, full-moon dates, and interesting facts.
http://moonphases.info/

Neatorama: A blog that features interesting stories from the world of science.
http://www.neatorama.com/

Null Hypothesis: Home of *The Journal of Unlikely Science,* this site is all about the tough questions in science.
http://www.null-hypothesis.co.uk/

Sandlot Science: This site features all sorts of optical illusions, some animated and some in JAVA applets.
http://www.sandlotscience.com/

Schlock Mercenary: On Schlock Mercenary, you can travel the galaxy to meet new and fascinating life forms.
http://www.schlockmercenary.com/

Science of Everyday Life: On this site, visitors can learn about the innovations that shape daily life.
http://scienceofeverydaylife.discoveryeducation.com/flash/iTimeline/itimeline.html

Sizing Up the Universe: This site offers a cool way to envision the outsize proportions of space.
http://www.smithsonianeducation.org/idealabs/universe/index.html

Snopes.com: This is the definitive Internet reference source for urban legends, folklore, myths, rumors, and misinformation.
http://www.snopes.com/

Sodaconstructor: A virtual construction kit for interactive creations using masses and springs.
http://sodaplay.com/creators/soda/items/constructor

Spiked Math: A comic all about math.
http://spikedmath.com/

The Internet Archive: Find out what a particular website looked like years ago by utilizing The Internet Archive.
http://archive.org/index.php

The Kitchen Pantry Scientist: Carry out science experiments in your kitchen with this website.
http://kitchenpantryscientist.com/

The Scale of the Universe 2: Explore the size and scale of objects in the universe with this fun interactive gadget.
http://htwins.net/scale2/

The Science Spot: Offers classroom activities for science teachers, including daily science trivia to help challenge students to learn.
http://sciencespot.net/

USBGeek: Dedicated to bringing the latest USB stuff, technology, and gadgets straight from the worldwide web to its customers.
http://www.usbgeek.com/

Whyville: Many games and activities are available on this site.
http://www.whyville.net/

Wonderville: Several tutorials, explanations, videos, activities, and more can be accessed on this website.
http://www.wonderville.com/

TALK TO ME!

Do you have a citizen science project you'd like people to know about? Are you a citizen scientist who would enjoy being featured on the site? Perhaps you'd just like to send a hello? Please do! Here's where I can be reached:

http://www.citizensciencecenter.com/get-in-touch/

I also accept guest posts (related to citizen science, of course).

http://www.citizensciencecenter.com/submit-a-guest-post/

ABOUT THE AUTHOR

My name is Chandra Clarke. My partner, Terence Johnson, and I founded Scribendi.com, an award-winning, ISO-certified editing and proofreading company; Scribendi was acquired in 2017. We currently own Inklyo.com, an online writing courses publisher. If the letters behind a name are important to you, I have a college diploma in industrial robotics, my BA was in English and Psychology, and my MSc was in Space Exploration Studies. I'm now doing a PhD in Creative Writing. There is a method to my apparent educational madness: the prevailing themes are understanding automation, innovation, the public understanding of science, communication, and motivation.

Endnotes

1 "History of the Christmas Bird Count," Audubon, accessed July 1, 2013, http://birds.audubon.org/history-christmas-bird-count.

2 Timothy Ferris, Seeing in the Dark: How Amateur Astronomers Are Discovering the Wonders of the Universe (New York: Simon & Schuster, 2003).

3 "Charles Darwin," BBC, accessed July 3, 2013, http://www.bbc.co.uk/history/historic_figures/darwin_charles.shtm

4 "Percival Lowell," Encyclopedia Britannica, accessed July 1, 2013, http://www.britannica.com/EBchecked/topic/349831/Percival-Lowell.

5 Ilona Miko, "Gregor Mendel and the Principles of Inheritance," Scitable, accessed July 3, 2013, http://www.nature.com/scitable/topicpage/gregor-mendel-and-the-principles-of-inheritance-593.

6 "Mary Elizabeth Barber," Darwin Correspondence Project, accessed July 7, 2013, https://www.darwinproject.ac.uk/person/namedef-271.

7 "Eva Ekeblad," Historiesajten, accessed July 7, 2013, http://www.historiesajten.se/visainfo.asp?id=582. See also "Eva Ekeblad," Wikipedia, accessed September 18, 2013, http://en.wikipedia.org/wiki/Eva_Ekeblad.

8 At the time of writing, Hostgator.com was offering packages ranging from US$4 to $10 per month.

9 Here I used http://www.comcast.com/internet-service.html, accessed July 18, 2013, as an example.

10 John C. McCallum, "Memory Prices (1957–2013)," accessed July 18, 2013, http://www.jcmit.com/memoryprice.htm.

11 "Open source" means that not only is the software free, but the source code, or programming behind the software, is accessible and can be modified.

12 oDesk, accessed July 18, 2013, https://www.odesk.com/o/profiles/browse/c/web-development/sc/web-programming/min/5/max/10/.

13 "More than 2 million books, several million articles, and thousands of special collections are scanned or born digital every year." "Exploring New Frontiers to Promote Digital Collections," WesternTrek, accessed July 21, 2013, contentdm.library.unr.edu/digitalprojects/westerntrek.pdf.

14 At http://www.AliBaba.com, for example, there are currently 135,488 different types of sensors available for purchase.

15 Kathryn Zickuhr, "Tablet Ownership 2013," Pew Internet, accessed July 21, 2013, http://www.pewinternet.org/Reports/2013/Tablet-Ownership-2013.aspx.

16 Aaron Smith, "Smartphone Ownership 2013," Pew Internet, accessed July 21, 2013, http://www.pewinternet.org/Reports/2013/Smartphone-Ownership-2013.aspx.

17 "FAQ," Hubble, accessed September 18, 2013,
http://www.spacetelescope.org/about/faq/.
18 More mind-blowing statistics: "When the Sloan Digital Sky Survey started
work in 2000, its telescope in New Mexico collected more data in its first few
weeks than had been amassed in the entire history of astronomy. Now, a
decade later, its archive contains 140 terabytes of information. A successor,
the Large Synoptic Survey Telescope, due to come on stream in Chile in 2016,
will acquire that quantity of data every five days" (emphasis mine). "Data,
Data Everywhere," The Economist, accessed July 21, 2013,
http://www.economist.com/node/15557443.
19 Mapper, accessed September 18, 2013, http://www.getmapper.com/.
20 If you really want the math: assuming a standard work week of 37.5
hours, and no holidays, vacations, or messing about on Facebook, this works
out to 59.3 straight weeks of image classification. What this would do for
your sanity is another question altogether.
21 Larry Greenemeier, "Forecast for Processing and Storing Ever-Expanding
Science Data: Cloudy," Scientific American, accessed July 21, 2013,
http://www.scientificamerican.com/article.cfm?id=science-data-cloud-
computing.
22 Brian Wang, "Google Using Artificial Intelligence to Improve Image and
Video Classification," Next Big Future, accessed July 23, 2013,
http://nextbigfuture.com/2013/03/google-using-artificial-intelligence-
to.html.
23 Clay Shirky, Cognitive Surplus: How Technology Makes Consumers into
Collaborators (New York: Penguin, 2010).
24 Ibid., 10.
25 Robert Lee Hotz, "When Gaming Is Good for You," The Wall Street
Journal, accessed July 23, 2013,
http://online.wsj.com/article/SB100014240529702034586045772632739431
83932.html.
26 Shirky, 20.
27 http://www.youtube.com/yt/press/statistics.html.
28 http://www.google.com/trends/.
29 "Keyword Planner Has Replaced Keyword Tool," Google, accessed July 23,
2013, https://adwords.google.com/select/KeywordToolExternal.
30 "Research and Development Expenditure (% of GDP)," Google, accessed
September 18, 2013,
http://www.google.com/publicdata/explore?ds=d5bncppjof8f9_&ctype=l&m
et_y=gb_xpd_rsdv_gd_zs.
31 Varvara Trachana, "Austerity-led Brain Drain Is Killing Greek Science,"
Nature, accessed July 23, 2013, http://www.nature.com/news/austerity-led-
brain-drain-is-killing-greek-science-1.12813.
32 Michele Catanzaro, "Spanish Scientists Protest to Save Research," Nature,
accessed August 1, 2013, http://www.nature.com/news/spanish-scientists-
protest-to-save-research-1.13207.

33 Jonathan Amos, "UK Science Spending to Remain 'Flat,'" BBC, accessed August 1, 2013, http://www.bbc.co.uk/news/science-environment-23065763.
34 Consider the infographic at http://www.scientificamerican.com/article.cfm?id=money-for-science, which demonstrates how, in the United States, defense spending dwarfs the outlay on both the National Science Foundation and NASA combined. Mark Fischetti, "Money for Science: U.S. Funding over the Years," Scientific American, accessed August 3, 2013.
35 "Poliomyelitis," World Health Organization, accessed August 3, 2013, 2013, http://www.who.int/mediacentre/factsheets/fs114/en/.
36 "Diarrheal Disease," World Health Organization, accessed September 18, 2013, http://www.who.int/mediacentre/factsheets/fs330/en/index.html
37 "Cancer Facts & Figures 2013," American Cancer Society, accessed August 3, 2013, 2013, http://www.cancer.org/research/cancerfactsstatistics/cancerfactsfigures2013/index.
38 Elliot Park, "Sense of Purpose Makes You Happy and Healthy," UCLA Magazine, accessed July 7, 2013, http://magazine.ucla.edu/exclusives/sense-of-purpose-makes-you-happy-and-healthy/.
39 Emi Kolawole, "A Lot of People Are Game for a One-Way Ticket to Mars—More than 100,000 of Them," Washington Post, accessed September 18, 2013, http://www.washingtonpost.com/blogs/innovations/wp/2013/08/12/a-lot-of-people-are-game-for-a-one-way-ticket-to-mars-more-than-100000-of-them/.
40 Not to mention how speculative that proposed mission to Mars sounds. However, as Washington Post blogger Emi Kolawole points out, "at the very least, it is a testament to how powerful the dream of space travel is for this planet's human inhabitants." Ibid.
41 Patrick Thibodeau, "Science-and-engineering Workforce Has Stalled in U.S., Report Says," Computerworld, accessed August 21, 2013, www.computerworld.com/s/article/9224823/Science_and_engineering_workforce_has_stalled_in_U.S._report_says.
42 United States Census 2010, accessed August 1, 2013, http://www.census.gov/2010census/.
43 You don't need the latest and fastest computer hardware to participate; indeed, if you've got some older units gathering dust, this might be a way to put them to work again before you recycle them for good. Just bear in mind that the slower and older the computer, the slower its work rate. If you plan to install the software on the computer you regularly use, and it's a slower model, you may want to adjust the settings of the software so that it works only when the computer is idle. Otherwise, the software may be set to work "in background," which might slow down the other applications you're trying to use.

44 Patrick Thibodeau, "China Surpassing U.S. with 54.9 Petaflop Supercomputer," Computerworld, accessed September 1, 2013, http://www.computerworld.com/s/article/9239710/China_surpassing_U.S._with_54.9_petaflop_supercomputer.
45 "Captain Ridley's Shooting Party," Bletchley Park, accessed September 8, 2013,
http://www.bletchleypark.org.uk/content/hist/worldwartwo/captridley.rhtm
.

46 This term is credited to Kevin Ashton. See Kevin Ashton, "That 'Internet of Things' Thing," RFID Journal, accessed September 18, 2013,
http://www.rfidjournal.com/articles/view?4986.

www.ingramcontent.com/pod-product-compliance
Lightning Source LLC
Chambersburg PA
CBHW071614170526
45166CB00003B/1082

9 781500 595500